SUPER COOL SPACE FACTS

SUPER COOL SPACE FACTS

A FUN, FACT-FILLED SPACE BOOK FOR KIDS

BRUCE BETTS, PhD

ILLUSTRATIONS BY STEVE MACK

ROCKRIDGE PRESS

For general information on our other products and services or to obtain technical support, please contact our Customer Care Department within the United States at (866) 744-2665, or outside the United States at (510) 253-0500.

Rockridge Press publishes its books in a variety of electronic and print formats. Some content that appears in print may not be available in electronic books, and vice versa.

TRADEMARKS: Rockridge Press and the Rockridge Press logo are trademarks or registered trademarks of Callisto Media Inc. and/or its affiliates, in the United States and other countries, and may not be used without written permission. All other trademarks are the property of their respective owners. Rockridge Press is not associated with any product or vendor mentioned in this book.

Cover and Interior Designer: Merideth Harte
Art Producer: Sara Feinstein
Editor: Susan Randol
Production Editor: Andrew Yackira
Illustrations © Steve Mack
Author photo courtesy of © Jennifer Vaughn

ISBN: Print 978-1-64152-521-3
eBook 978-1-64152-522-0

For my sons, KEVIN and DANIEL

CONTENTS

WELCOME TO OUTER SPACE!

Join me for a fun, fact-filled adventure through outer space. We'll explore *stars*, *planets*, rockets, astronauts, and many more amazing things. This is a book about *space* for kids. It is written for kids ages four to eight, but space facts are fun for everyone.

I love space so much that I became a scientist who studies planets—a planetary scientist. Then I realized I really love teaching others about space. The objects in space are incredible. They are big and weird and just plain cool. There are huge, hot stars, like the Sun, and planets with giant volcanoes and canyons, and *moons* with oceans under ice. I can go on and on about space and how cool it is. That is why I wrote this book—to share the joy of space with you. I will be your guide as we go on a space adventure.

Space is also exciting because although we know a lot about it, there are still lots of mysteries to solve. To learn about space, scientists use telescopes and *spacecraft*. We are making new discoveries all the time.

In our trip through space, we will start big—really big. We'll talk about the *universe*, all the stuff that is out there, your place in the universe, and how we learn about things in space. Then we'll talk about different kinds of stars, including the strange things they leave behind when they explode. After that, we'll come closer to home and learn about our *Solar System*, including the Sun and planets. I'll also tell you about objects you can see in the sky. We'll finish our tour by talking about rockets and spacecraft and astronauts and your space future.

In this book you'll find a basic description of each thing we talk about, some awesome pictures, and tons of super cool space facts. And jokes—lots of jokes. There is a glossary in the back where you can find definitions of space words used in this book. (Words that appear in the glossary are in pink the first time they're used.)

Did you know you're already a space traveler? You are! You are on a planet called Earth that travels through space around a star named the Sun. Are you ready to travel to other places in space? I am.

LET'S GO!

THE UNIVERSE

The universe is all of space and everything in it. That includes Earth, the farthest stars from Earth, and everything in between. We've learned a lot about what is out there in the universe, what things are made of, and even how old the universe is. But there is much we do not know, which is part of what makes space so interesting. Before we get to the big topic of the big universe, though, let's talk about your place in it.

YOUR PLACE IN THE UNIVERSE

Where are you in the universe? You live on a planet called Earth. Earth goes around a star called the Sun. Earth rotates—in other words, it spins like a basketball can spin on someone's finger.

When you're on the side of Earth that's facing the Sun, it is daytime. When you're on the side of Earth that's facing away from the Sun, it is nighttime.

A picture of Earth taken from space. You live on Earth.

Earth goes all the way around the Sun in one year. Our year is about 365 days long.

The Sun appears much bigger and much brighter than the other stars because the Sun is much closer to us.

Planets and other objects that go around the Sun are part of what we call our Solar System. There are many solar systems in the universe.

Our Solar System is part of the Milky Way Galaxy, which is a very big group of stars. There are many galaxies in the universe.

Just like your house has an address, your address in the universe would be Earth, the Solar System, the Milky Way Galaxy. That is your place in space.

Earth rotates once in a day—in other words, once every 24 hours.

You cannot see other stars in the daytime because the Sun's light is very bright and it bounces around in Earth's *atmosphere* (the air that surrounds Earth), overwhelming the stars.

3

ROBBY: "WE'RE GOING TO SEND A SPACECRAFT TO THE SUN."

BOBBI: "HOW CAN YOU DO THAT? THE SUN IS TOO HOT."

ROBBY: "WE'LL SEND IT AT NIGHT!" 🙂

4

SPACE & THE UNIVERSE

Space, sometimes called outer space, is all the stuff beyond Earth and its atmosphere. Most of space has almost nothing in it. There will be a little something here and there, and occasionally there will be a planet or moon or star, but mostly space is empty.

The objects that do float around in space are very interesting. Much of what is out there is in groups of stuff like solar systems and galaxies.

Many things go around other things. The path an object follows is called an *orbit*. Moons, like Earth's Moon, orbit planets. Planets are big, ball-shaped objects that orbit stars. Stars often orbit the center of a *galaxy*.

EARTH

VENUS

MERCURY

SUN

MARS

The orbits (paths) of the four planets closest to the Sun, which is the pink dot at the center.

Many orbits look like circles. But some look like squished circles called ellipses.

Some planets are rocky, like Earth. Some planets are made of mostly gas, like Jupiter.

What keeps Earth from flying away from the Sun? What keeps a football from flying away from Earth? It is the same thing that keeps you on the ground. It is also what causes something to fall when you drop it. It is a force called gravity. All objects create gravity, but only very massive (heavy) objects like planets and stars have lots of gravity.

How do we know what's out there in space if everything is so far away? We use spacecraft to explore our Solar System. (Everything else is too far away to reach with spacecraft.) We also use telescopes to study objects in our Solar System as well as objects that are far away.

The Moon has less gravity than Earth. That means you can jump higher on the Moon than you can on Earth.

WHAT IS GRAVITY'S FAVORITE SEASON?

FALL.

A large telescope: the Palomar 200 inch.

TELESCOPES

Telescopes help us study objects that are far away. When you look through a telescope, what you see looks bigger and closer. Astronomers use telescopes to study things in space, and telescopes with cameras can take pictures of those objects. Scientists use very large telescopes to study things that are super far away.

Part of the inside of a huge telescope, the northern Gemini telescope. Some of its 8 meter mirror can be seen at the bottom.

Some telescopes use lenses, kind of like what are used in eyeglasses. Some telescopes use curved mirrors, and some use both lenses and mirrors.

Astronomers sometimes use telescopes to look at light that your eye cannot see, like ultraviolet and infrared light.

Astronomers put their big telescopes on top of mountains. That way they are above many of the clouds and other effects of the atmosphere that make it hard to see faraway objects.

WHY DID THE TELESCOPE HAVE TROUBLE DOING HOMEWORK?

IT COULDN'T FOCUS. ☺

SPACE TELESCOPES

Some telescopes have been launched into space. Though it is hard and expensive to make things work in space, it is worth the effort. Space telescopes are completely above clouds, hazes, and all of Earth's atmosphere. That means they can see really well. Some of the most spectacular images of our universe have been produced by space telescopes.

The Hubble Space Telescope photographed from the space shuttle Discovery.

If you could see as well as the Hubble Space Telescope, you could distinguish between two fireflies next to each other on the other side of the United States.

The Hubble Space Telescope has been working in space since 1990.

You may know X-rays are used to see bones in people's bodies. Astronomers use X-ray space telescopes like the Chandra X-ray Observatory to study stars and other space stuff.

HOW DO WE KNOW WHAT DISTANT THINGS ARE MADE OF?

Objects in space are really far away. We have sent spacecraft to only a few places. So how do astronomers know what things are made of?

They use patterns in the colors of light to figure out what they're seeing. Here is an Earth example: Green often means something is a plant. Brown is usually the color of dirt. But brown also is often the color of tree trunks. How would an astronomer tell dirt from a tree trunk?

She would use scientific instruments that carefully split the light into different colors, like a rainbow. Then she would measure how bright each part of each color is. And she might use instruments that can see light our eyes can't, like infrared or ultraviolet light.

A picture from Mars taken by a robotic rover named Spirit.

Mars looks red because of rusty rocks. The iron in the rocks has rusted, just like iron on Earth can rust.

WHAT IS HELIUM'S FAVORITE DESSERT?

AN ICE CREAM FLOAT.

Different materials absorb or reflect different colors of light. The astronomer would look for patterns of colors, since patterns tell what things are made of. Just as every person has different fingerprints, every object has different patterns in its colors of light.

BIG WORD ALERT

Using the detailed colors of light to learn what things are made of is called *spectroscopy*.

Using spectroscopy, we learned that stars, including the Sun, are mostly made of hydrogen, the lightest element.

Astronomers found helium on the Sun before it was discovered on Earth. They found a pattern they did not recognize in the color details of the Sun's light. We later discovered helium on Earth. Helium is what fills balloons that float.

Light from the Sun split into a rainbow. The black lines can tell us what the Sun is made of.

THE BIG BANG

All the galaxies, or groups of stars, are moving away from us and away from one another. It is kind of like inflating a balloon. You can see that all the spots on the balloon get farther away from one another as you blow it up.

Now, imagine going back in time. All those galaxies will get closer and closer to one another. If you go back in time far enough, everything in the universe will be in one tiny space.

Thinking like this led to the Big Bang theory. The universe started with a "Big Bang." It was like the biggest explosion ever! Ever since the Big Bang, objects have been moving apart from one another.

Hubble Space Telescope picture of a tiny part of the sky showing hundreds of galaxies.

WHY DID THE FIREWORK REMIND JANE OF THE UNIVERSE?

BECAUSE OF THE BIG BANG. ☺

The Big Bang occurred 13.8 billion years ago—that is 13,800,000,000 years ago. That's how old the universe is.

What happened before the Big Bang? We don't know. We do know the Big Bang theory does a great job explaining what we see in the universe now.

The Hubble Space Telescope is named after an astronomer named Edwin Hubble. Through careful telescope measurements, he confirmed the expansion of the universe.

GALAXIES

A galaxy is a group of many, many stars, as well as dust and gas. Galaxies can be seen through telescopes. A galaxy is held together by gravity.

There are three basic types of galaxies. *Spiral galaxies* are named for their spiral shape. They are kind of flat—sort of shaped like a plate. *Elliptical galaxies* look like blobs with no distinct shapes in them. And *irregular galaxies* are anything else. We live in a spiral galaxy called the Milky Way Galaxy.

12

A galaxy can have millions, billions, or even trillions of stars.

A telescopic picture of the spiral galaxy Andromeda.

WHAT IS THE ASTRONOMER'S FAVORITE SOCCER TEAM?

THE LOS ANGELES GALAXY. 😊

The Andromeda Galaxy can be seen with just your eyes. It is a fuzzy-looking blob. You need to know where to look as well as have clear, dark skies.

Astronomers think there are hundreds of billions of galaxies in the universe.

THE MILKY WAY

We live in the Milky Way Galaxy. Our Sun is one of many stars in the Milky Way. The Milky Way is a spiral galaxy. We are inside the Milky Way. We are about halfway between the center of the galaxy and the outer edge. Everything in our galaxy, including Earth and our Solar System, revolves around the center of the galaxy. It takes a very long time for our Solar System to move all the way around our galaxy (about 250,000,000 years).

The Milky Way wasn't named after a candy bar; the candy bar was named after our galaxy.

A picture of the sky showing bright stars and dark dust of the Milky Way Galaxy.

13

Continued >

THE UNIVERSE

MILKY WAY

Because the Milky Way is mostly flat, and we are inside it, we see the Milky Way's big groups of stars as a cloudy stripe across the sky. You can see it if you are away from city lights. That stripe looks "milky." In other words, it has a fuzzy white brightness.

14

That fuzzy stripe was named the Milky Way before people knew we live in a galaxy.

Another picture of the milky stripe across the sky made up of the stars of the Milky Way.

All the individual stars you see with just your eyes are in the Milky Way Galaxy.

WHAT IS THE ASTRONOMER'S FAVORITE CANDY BAR?

A MILKY WAY. 🙂

NEBULA

A *nebula* is a giant collection of dust and gas in space. Nebulae can look very pretty in telescopes, particularly in pictures taken by large telescopes. Some nebulae are left over from material blown out at the end of stars' lives. Other nebulae are places where new stars are being born.

The Orion Nebula is visible with just your eyes. It looks like a fuzzy patch in the middle of the "sword" in the *constellation* Orion.

The singular of *nebulae* is *nebula*. So we say one nebula, but two nebulae.

15

A Hubble Space Telescope image of the Orion Nebula.

Continued >

NEBULA

The Crab Nebula is leftover gas and dust from a star explosion (*supernova*).

The Horsehead Nebula actually looks like a horse's head.

WHY DID THE STREET SWEEPER DISLIKE THE NEBULA?

BECAUSE OF ALL THE DUST. 🙂

A space telescope picture of the center of the Lagoon Nebula showing dark dust and hazy gas clouds.

DARK MATTER

What is dark matter? We don't know. It is one of the mysteries of modern astronomy. Think of matter as stuff. Astronomers use the term *dark* to mean they can't see that stuff with telescopes. Why do they think something is there if they can't see it? Because to explain lots of things they do see, there has to be more stuff than what they are seeing.

A picture of the Whirlpool Galaxy.

Most of the matter in the universe is thought to be dark matter.

The additional gravity from dark matter would explain why galaxies don't fly apart.

DANIEL: "DO YOU LIKE MY COSTUME?"

KEVIN: "I CAN'T SEE YOU."

DANIEL: "EXACTLY! I'M DRESSED AS DARK MATTER." ☺

DARK ENERGY

Just like dark matter, we don't know what dark energy is. It is another mystery of modern astronomy. Since the Big Bang, stuff in the universe has been getting farther apart. This is true of galaxies today. Astronomers have discovered that things are getting farther apart faster with time. The expansion of the universe is accelerating. We don't know why. Astronomers named whatever is causing this dark energy.

A Hubble Space Telescope picture showing lots of galaxies.

Dark energy was named in 1998 and was named to be like the term *dark matter.*

Dark energy is currently thought to make up about 68 percent of the energy in the universe.

LIFE OUT THERE?

Is there life out there beyond Earth? Are there aliens? As of now, no one has found aliens or life beyond Earth. There is no evidence of either, but that doesn't mean they don't exist. The universe is a very big place.

Sometimes radio telescopes are used to try to listen for alien signals.

Three things are required by all life as we know it:

1. A source of energy (for example, sunlight)
2. Certain elements, like carbon
3. Liquid water

Scientists are searching for messages from aliens by searching for radio signals and laser signals. They have not found any.

WHAT IS AN ALIEN'S FAVORITE THING ABOUT TEATIME?

THE SAUCER. 🙂

Continued >

LIFE OUT THERE

Our Solar System has planets and moons that could have life. If life exists in our Solar System, it is probably microbial life, like bacteria. But there are many planets in other solar systems, and some of them will be similar to Earth. There could be intelligent life on them. We just don't have evidence of that . . . yet.

What are some places in the Solar System that have—or had—liquid water (so they could have—or had—life)?

Mars, especially in its past, when it had lots of liquid water

Europa, a moon of Jupiter—astronomers think it has a liquid water ocean under ice

Enceladus, a moon of Saturn, may have a liquid water ocean under ice and has water geysers

CHAPTER 2

STARS AND CONSTELLATIONS

Stars are fun. They fill the night sky. They form patterns and shapes in the sky that you can learn to recognize. They look small because they are very far away, but they are actually huge and hot. The Sun is a star that is much closer to us. Stars come in different sizes, colors, and brightness. Sometimes they explode at the end of their lives and leave behind something really weird.

STARS

Stars are giant balls of hot glowing gas. The Sun is a star. All stars except the Sun are so far away that they appear as dots of light in the night sky. If you look carefully at stars in the night sky, you can see some are red. Those stars are cooler than white stars, which are cooler than blue stars. But they are all extremely hot.

22

Hundreds of stars make up the Quintuplet star cluster.

WHAT DID THE COOL STAR SAY TO THE HOT STAR?

HEY, WHY SO BLUE? 😊

Stars come in different sizes, but we can't see that with just our eyes because they are so far away.

Some solar systems, like ours, have one star. Some have two or more stars that orbit one another.

Some of the stars in the Hyades star cluster. Note their different colors.

BIG WORD ALERT

Technically, stars are made of *plasma*, which is like an electrical charge–filled gas.

STARS AND CONSTELLATIONS

CONSTELLATIONS

Stars form shapes and patterns in the sky. Some of these patterns are called constellations—patterns that astronomers everywhere have agreed upon. Constellations are often named after what people thought they looked like. An example is Canis Major, which in Latin means "Greater Dog." Another is Scorpius the Scorpion. Some constellations are easy to recognize, like Orion. Others are trickier, but you can learn to identify them all.

The constellation Orion. The three stars in a line are called Orion's Belt.

There are 88 constellations.

Astronomers use the 88 constellations to divide up the sky. It is kind of like how country boundaries divide up Earth's land.

Some constellations are named after creatures from mythology. For example, the largest constellation is Hydra, named after a snake-like creature from Greek mythology.

Orion on the right and Canis Major (Greater Dog) on the left. Do you think it looks like a dog?

ASTERISMS

An *asterism* is any group of stars that makes a pattern in the sky. Probably the most famous example is the Big Dipper. The Big Dipper consists of seven stars that look like a "dipper"—a ladle or spoon. Asterisms are usually easy to recognize and contain bright stars.

The Big Dipper asterism.

The asterism called the Big Dipper in the United States is often called the Plough in the United Kingdom.

There is also a Little Dipper asterism.

The Big Dipper is part of the bigger constellation Ursa Major, which means "Greater Bear."

How can you tell the difference between a constellation and an asterism? Constellations refer to a part of the sky, not just the stars. And there are 88 constellations all astronomers have agreed to use—asterisms are any other pattern of stars.

The Big Dipper can be used to find the North Star, which is always in the direction North.

There is a Northern Cross asterism and a Southern Cross asterism. Both look like crosses.

WHAT DID THE ASTRONOMER SAY TO THE LEAKY CUP?

YOU'RE A BIG DRIPPER. 😊

The Summer Triangle asterism formed by three bright stars. Can you find the Northern Cross?

STAR CLUSTERS

A *star cluster* is a group of stars. They are near one another in the sky. They are also near one another in space. They are usually all about the same age.

There are two different kinds of star clusters. *Open clusters* usually have a few hundred young stars. *Globular clusters* usually have a few hundred thousand old stars.

A Hubble Space Telescope picture of the core of the Great Globular Cluster.

Some clusters, like the Pleiades, are visible with just your eyes. Many more are visible with binoculars.

WHAT IS THE ASTRONOMER'S FAVORITE CEREAL?

STAR CLUSTERS. 🙂

SUPER COOL SPACE FACTS

Some clusters have been given silly names based on what they look like, including the Butterfly Cluster, Beehive Cluster, and Wild Duck Cluster.

A picture of part of the Beehive Cluster.

LIFE CYCLE OF STARS

Stars are born from huge clouds of gas and dust in space. They aren't born like people, of course. The gas and dust fall together under gravity, heat up, and a star is born.

Stars, including the Sun, burn for a very long time. This is the longest part of their lives. The Sun has been burning for almost five billion (5,000,000,000) years. It will keep burning for at least five billion years more.

The Pleiades star cluster, with new stars among clouds of dust and gas.

BIG WORD ALERT

What a star turns into is determined not by how big it is, but by how massive it is—in other words, how much *mass* it has. Mass is like weight, but it doesn't factor in gravity.

Some nebulae are star-forming regions. The Pleiades star cluster is an example. Dust is coming together there, forming cute little baby stars.

When stars reach the end of their lives, weird things happen. They turn into things like *red giants*, *white dwarfs*, and black holes. What they turn into is determined by how massive they are.

BIG WORD ALERT

Stars don't actually burn like a fire. They convert hydrogen to helium through a process called *nuclear fusion*.

Hotter stars don't live as long as cooler stars (though cooler stars are still hot). The Sun is a middle-temperature star.

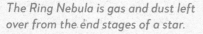

The Ring Nebula is gas and dust left over from the end stages of a star.

31

STARS AND CONSTELLATIONS

RED GIANTS

Most stars turn into red giants near the end of their lives. The star expands and can become enormous. It looks red but still like a dot because it is so far away. With just our eyes, we can see some red stars in the sky that are red giants.

32

The bright star Betelgeuse in Orion is a red giant. It is near the top of this picture.

When a star gets bigger and becomes a giant, its surface gets cooler. That causes the red color.

Examples of bright red giants include Betelgeuse in Orion, Antares in Scorpius, and Aldebaran in Taurus.

Red giants can be huge. Some of them are so big that if they were in our Solar System, they would be bigger than the orbit of Mars. That is more than 300 times the diameter of our Sun.

NOVA AND SUPERNOVA

A nova is a star that suddenly gets brighter. A supernova is a star that suddenly gets much, much brighter. A supernova happens when a huge star explodes at the end of its life. Supernova explosions are gigantic! A supernova will fade over weeks and months and years.

A nova usually results from the interaction of two stars. A supernova usually results from an exploding star.

A supernova can temporarily be brighter than a whole galaxy.

A supernova leaves behind a dust and gas nebula.

This telescope picture shows a supernova to the lower left of a galaxy.

WHITE DWARF

A white dwarf is what is left at the end of the lives of most stars. It forms when the star runs out of fuel to burn. This is often the stage after a red giant.

White dwarfs are much smaller than stars. That is why they are called dwarfs. They are about the size of Earth.

A white dwarf still has the mass (like weight) of a star. But it is the size of a planet. It is very dense!

This telescope picture shows the bright star Sirius A. The dot to its lower left is its white dwarf companion, Sirius B.

NEUTRON STARS

After really big stars explode, they leave behind *neutron stars*. Neutron stars are very small. They are about the size of a city. But they still have the mass of a star. They are super dense and spin very fast.

Telescope picture of a neutron star. The arrow is pointing to it.

Neutron stars have about the density you'd get by squishing all the humans on Earth into the size of a sugar cube.

Because they are so small, few neutron stars have been seen with telescopes.

Why do neutron stars spin so fast? When a figure skater pulls their arms in, they spin faster. Similarly, when a big star collapses into a tiny neutron star, it spins much faster.

WHY DIDN'T THE NEUTRON STAR GET THE JOKE?

BECAUSE IT WAS REALLY DENSE. 😊

STARS AND CONSTELLATIONS

PULSARS

Pulsars are a kind of neutron star that spins. Neutron stars that put out beams of light are called pulsars. Astronomers with the right kind of telescope can see pulses or flashes of light from pulsars. The spinning pulsar is like a lighthouse or someone spinning with a flashlight. You see the light each time it points at you.

The Vela pulsar seen in X-rays. It is the bright dot at the center and it is spewing out stuff.

When a pulsar was first discovered, people wondered whether the signal was evidence of aliens. Then they found other pulsars in other parts of the sky. Scientists figured out the signals were from neutron stars.

Pulsars spin fast. Some spin once a second. The fastest one spins 716 times per second!

BLACK HOLES

You may have heard of black holes. They are super weird. A really, really huge (massive) star will form a black hole at the end of its life. This happens after a supernova. Only the biggest (most massive) stars will keep shrinking past a white dwarf, past a neutron star, and into a black hole. They have so much gravity that even light cannot get out of a black hole. That is why we call them black.

We can't see any light from black holes. Sometimes we can see X-rays or other light coming from objects as they enter a black hole.

Telescope view of the center of our Milky Way Galaxy. There is a black hole there that you can't see.

Continued >

Astronomers think most galaxies have black holes at their center.

The first picture of a black hole is of the amazingly huge black hole at the center of a galaxy called M87. The black hole has a mass equal to 6.5 billion (6,500,000,000) Suns.

The first picture of a black hole and its "shadow," which appear as the dark circle in the middle of the bright glowing gas ring.

There are no black holes anywhere near us.

EXOPLANETS

Planets like Earth and Mars orbit the Sun. There are also planets that orbit other stars. We call them exoplanets. We have discovered thousands of them. Some are big like Jupiter. Some are smaller than Earth.

An artist's painting of an imagined exoplanet and its star.

Exoplanets are very hard to see. Astronomers haven't seen very many. Most are discovered by studying light from their stars. For example, we can see the starlight dim when a planet passes in front of the star.

The first exoplanet discoveries happened in the 1990s.

Some exoplanet solar systems look very different from ours. Some have big planets near their star. Our big planets are far from the Sun.

THE SOLAR SYSTEM

We live in *the* Solar System. The Solar System consists of the Sun and all the planets and other objects that orbit the Sun. That includes Earth. It also includes really hot planets and really cold planets and really cold planets, giant planets and *dwarf planets*. Some planets and moons have mountains and volcanoes, while others have hurricane-like storms that are bigger than Earth. Let's explore our weird and interesting Solar System. We'll start with a look at the whole Solar System, then talk about its biggest object: the Sun. After that, we'll move out from the Sun toward the cold, distant parts of our Solar System.

THE SOLAR SYSTEM

The Solar System is our home star system. You can think of the Sun as the center of the Solar System. Everything else in the Solar System orbits the Sun.

There are eight planets. The four closest to the Sun are rocky. That includes Earth. The four farthest from the Sun are much bigger and they are not rocky. They are huge balls of gas. They do *not* have surfaces you could stand on.

Moons go around planets or other objects. Earth has one moon we call *the* Moon. Some planets have lots of moons.

There are many smaller objects orbiting the Sun, including rocky *asteroids* and icy *comets*.

The Solar System formed more than four billion years ago. It formed from a cloud of gas and dust.

The Solar System is big. If you could drive a car straight to the Sun from Neptune (the planet farthest from the Sun), it would take over 5,000 years.

Almost all the mass in the Solar System is in the Sun—over 99.8 percent. So all the planets, moons, and other objects in the Solar System are really tiny compared to the Sun.

SUN

Mercury

Earth

Saturn

Uranus

Neptune

Venus

Mars

Jupiter

43

WHAT DID THE SUN ASK EARTH
TO DO FOR THE PARTY?

PLAN-ET. 🌚

If Earth were the size
of a baseball, Neptune
would be about the
size of a basketball,
and Jupiter the size
of a big, inflatable
exercise ball.

THE SUN

The Sun is a star. It is our star. It is hot and really, really big. The Sun is important to life.

The Sun is mostly made of hydrogen and helium plasma, which is sort of like an electrically charged gas. Deep inside the Sun, hydrogen is turned into helium. That process, called nuclear fusion, releases energy. That makes the Sun hot.

A picture of the Sun taken using a safety filter. The dark spots are called sun spots.

Though people often think of the Sun as yellow or orange, it actually is white. Earth's atmosphere can make it look yellow, orange, or even red.

The Sun is huge. More than one million (1,000,000) Earths could fit inside the Sun.

The Sun is bigger than the distance from Earth to the Moon.

The Sun provides the energy needed by most life on Earth. Light from the Sun is used by plants to grow. Animals eat those plants to get energy.

WARNING: NEVER LOOK AT THE SUN. LOOKING AT THE SUN COULD HURT YOUR EYES.

The size of Earth compared to the Sun is like the size of a mouse compared to an elephant.

WHY DID THE SUN GO TO SCHOOL?

TO GET BRIGHTER. ☺

A spacecraft picture of the Sun taken in ultraviolet light.

MERCURY

Mercury is the closest planet to the Sun. Mercury is the smallest planet. It is rocky and has almost no atmosphere (so it doesn't have air like Earth).

Mercury looks kind of like Earth's Moon. Mercury is covered in impact craters shaped like bowls. Craters are caused by space rocks hitting the ground at high speeds.

A spacecraft picture of Mercury.

Despite being the closest planet to the Sun, Mercury has water ice in the bottom of bowl-shaped craters near its poles. The bottoms of those craters never see sunlight, and there is no atmosphere (gas) to spread heat around.

Mercury is the speediest planet. At its fastest, it orbits the Sun at nearly 60 kilometers per second (134,000 miles per hour).

Mercury does not have a moon.

If you were standing on Mercury, the Sun would look three times wider than it looks from Earth.

The mountains on Mercury are named after the word for "hot" in various languages.

Mercury is very hot during its long daytime. Daytime lasts 88 Earth days. It is very cold during its long night, which is also 88 Earth days long. Daytime temperatures on Mercury are hundreds of degrees hotter than nighttime temperatures. During the day, it is hotter than an oven. During the night, it is much colder than a freezer.

A spacecraft picture of Mercury. Notice the many craters.

VENUS

Imagine a dangerous planet. Give it a surface hotter than an oven. Now add acid rain. Venus actually has those things.

Venus is the second planet from the Sun. It is almost as big as Earth. It is rocky. Venus spins the opposite way compared to Earth.

Venus has a much thicker atmosphere than Earth. That atmosphere acts like a blanket or a greenhouse—it traps heat. The surface of Venus is very hot. It is even hotter than Mercury. The surface stays hot day and night.

A spacecraft picture of Venus.

Venus has clouds and rain made of sulfuric acid.

Venus's atmosphere is mostly made of carbon dioxide. That is the same gas that makes bubbles in soda.

Venus has clouds all around the planet. They keep us from seeing the surface with normal cameras. Spacecraft have used radar to look through the clouds.

Venus is sometimes visible in the night sky. It will look like the brightest star in the night sky.

Venus does not have a moon.

Venus has so much gas in its atmosphere that it is kind of like being under the ocean. But there is no water, only gas. The surface pressure on Venus is like being 1 kilometer (3,000 feet) under Earth's ocean. Many submarines on Earth would be crushed on Venus's surface.

The surface of Venus is much hotter than a kitchen oven. Venus is almost 500 degrees Celsius (almost 900 degrees Fahrenheit).

Radar image of Venus with false color.

EARTH

Earth is our home planet. It is where we live. Earth is the third planet from the Sun. It is the biggest of the four rocky planets.

Earth is special. It has life—plants and animals (including humans). It is the only place we know of in the Solar System that has life.

Earth is sometimes called the Goldilocks Planet (from the story of Goldilocks and the Three Bears). Why? Well, Earth is just the right distance from the Sun. It is not too hot. It is not too cold. It is just right for life. It also has a Goldilocks atmosphere for life. Venus has too much atmosphere. Mars has too little. Earth has just the right amount.

A spacecraft picture of Earth showing North and South America, oceans, and clouds.

Earth's atmosphere is mostly made of a gas called nitrogen. It also has a lot of oxygen. We breathe air to get the oxygen. Earth has a lot of oxygen in its atmosphere because of plants. Plants give off oxygen.

Because of its Goldilocks temperature and atmosphere, Earth is the only place in our Solar System that has liquid water on its surface. All life on Earth needs liquid water.

Earth has one moon, called the Moon.

Most of Earth's surface is covered with oceans. More than two-thirds of the surface is covered with water.

From space, Earth looks mostly blue because of the oceans and white because of clouds.

Light is the fastest thing in the universe. Earth is so far away from the Sun, it still takes the Sun's light more than eight minutes to reach Earth.

THE MOON

The Moon is like Earth's friend in space. It travels with us around the Sun. It also travels around Earth. The Moon is much smaller than Earth, but it is one of the largest moons in the Solar System. The Moon is rocky and has almost no atmosphere.

You can see the Moon at night and even sometimes in the daytime. The Moon is close enough that we see it as a circle and not just a dot.

A picture of the Full Moon.

Apollo 15 astronaut Dave Scott on the Moon in 1971.

People used to think the dark areas on the Moon were seas. They are actually dark rocks.

The Moon's orbit is getting farther from Earth at about the same rate your nails grow—about 38 millimeters (1.5 inches) per year.

About 30 Earths could fit between Earth and the Moon.

Like Mercury, the Moon is covered in impact craters shaped like bowls. They are caused by space rocks hitting the ground at high speeds.

From Earth, we always see the same side of the Moon. We refer to that side as the Near Side. Earth's gravity and related tidal effects have caused this phenomenon.

Almost 100 missions have been flown or attempted to be flown to the Moon, including robotic flybys, *orbiters*, and *landers*.

The Moon is the only other place humans have walked besides Earth. Twelve astronauts walked on the Moon, all between 1969 and 1972.

A spacecraft picture of the Moon's surface showing many impact craters.

WHAT IS A COW'S FAVORITE PLACE IN THE SOLAR SYSTEM?

THE MOOOOOOOON. 😊

THE SOLAR SYSTEM

MARS

Mars is a rocky planet that is smaller than Earth. It is the fourth planet from the Sun. Mars is colder than Earth. It is called the Red Planet because of its color, which is caused by rusty red rocks and dust on its surface.

Mars has mountains, canyons, sand dunes, and plains. It looks like some deserts on Earth. It also has impact craters.

Mars has a very thin atmosphere, so it does not have much gas surrounding it.

Sometimes you can see Mars in the night sky, where it looks like a red or orange star.

Many robotic spacecraft are studying Mars. Some are going around it. Some are even driving around on the surface.

A spacecraft picture of Mars. Many cloud-free pictures were combined to create this.

The largest canyon system on Mars is huge. Called Valles Marineris, it would stretch all the way across the United States.

The surface of Mars is about the size of all of Earth's land (not including Earth's oceans).

The largest mountain in the Solar System is on Mars. Called Olympus Mons, it is a volcano more than twice as tall as Mt. Everest, Earth's tallest mountain.

The Curiosity rover on Mars next to a sand dune. This is a selfie the rover took.

Like Earth, Mars has icy polar caps. Like Earth's, they contain water ice. But the Mars polar caps also get covered by super cold carbon dioxide ice. That is what we call dry ice on Earth.

WHAT DO YOU CALL A SANDWICH MADE OF GRAHAM CRACKERS, MARSHMALLOW, AND A PLANET?

S'MARS. 😊

THE SOLAR SYSTEM

ASTEROIDS

An asteroid is a small rocky or metallic object. (At least it is small compared to a planet.) Between Mars and Jupiter is the Asteroid Belt. That is where millions of asteroids orbit the Sun.

The largest known asteroid is Ceres. Most asteroids, including Ceres, are located in the Asteroid Belt, though some asteroids come closer to the Sun.

Movies like *Star Wars* often make an asteroid belt look crowded. They are actually mostly empty. There are lots of asteroids, but they are spread out in a huge volume of space.

A spacecraft picture of the very rocky asteroid Bennu, which is five football fields wide.

WHAT DID ONE ASTEROID SAY TO THE OTHER ASTEROID?

YOU ROCK! ☺

The NASA Dawn spacecraft orbited two of the largest asteroids, Ceres and Vesta.

Ceres has almost one-quarter the mass of all the asteroids combined.

Ceres is both big and small. It is the biggest asteroid by far—at almost 1,000 kilometers (600 miles) wide—but it is much smaller than Earth and even Earth's Moon.

If you squished all the asteroids together, they would still be smaller than Earth's Moon.

A spacecraft picture of the asteroid Vesta, the second-largest asteroid. It is more than 500 kilometers (300 miles) wide.

JUPITER

Jupiter is the fifth planet from the Sun and the largest planet. It is one of the four planets called giant planets. It is made mostly of gas (so it is a gas giant). There is no solid surface, so you could not stand on Jupiter.

Jupiter has different colored clouds. They look like stripes across Jupiter. It also has huge storms. The biggest storm, which is like a giant hurricane, is red. It is called the Great Red Spot.

Jupiter is sometimes visible in the night sky. It will look like an extremely bright star.

A Hubble Space Telescope picture of Jupiter.

Jupiter spins faster than any other planet. Its day is about 10 hours long.

Jupiter's moon Europa has a liquid water ocean underneath lots of ice. This ocean has about twice as much water as all of Earth's oceans.

If Earth were the size of a soccer ball, then Jupiter would be about the height of a professional soccer goal.

Jupiter has 79 known moons. We continue to discover more. Most of the moons are small. But four are large and very interesting. Io has more volcanoes active than anywhere else in the Solar System. Europa has an icy surface and an ocean deep under the surface. Ganymede is the largest moon in the Solar System—it is bigger than Mercury. Callisto has an old cratered surface.

Jupiter and its moons are very cold. That's because they are far from the Sun. As we go farther from the warmth of the Sun, it gets colder. We find a lot of ice at Jupiter and beyond.

The Great Red Spot is bigger than Earth. It has existed for more than 300 years since its discovery.

Jupiter is so big that all the other planets in our Solar System would easily fit inside it.

Io is sometimes called the pizza moon. It looks a little like a pizza from space.

KNOCK, KNOCK.
WHO'S THERE?
I.
I WHO?
IO YOU FIVE DOLLARS. 😊

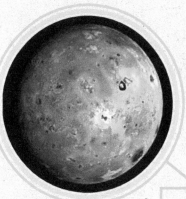

Io's volcanoes give it an unusual look.

SATURN

Saturn is the sixth planet from the Sun. It is smaller than Jupiter but still huge. It is a gas giant, so you could not stand on Saturn. Saturn is kind of yellow.

Saturn has amazing rings around it. They make Saturn look different from the other planets. Saturn's rings are made of dirty snowballs. There are billions of these snowballs. Some are the size of dust and some are as big as houses.

Saturn has 62 known moons. We continue to discover more. The moon named Titan is much bigger than the others. Titan is the only moon with a thick atmosphere. In other words, it is the only moon with a lot of gas surrounding it.

A spacecraft picture of Saturn and its rings.

Titan has lakes and seas, but they are not made of water. The lakes and seas are made of methane and ethane—what we call natural gas on Earth.

WHAT ARE SATURN'S FAVORITE MOVIES?

THE LORD OF THE RINGS. ☺

SUPER COOL SPACE FACTS

Most of Saturn's moons are covered in water ice. One of those moons is Enceladus. It has geysers that spit water ice into space.

Saturn is sometimes visible in the night sky. It will look like a bright yellow star.

You could fit 764 Earths inside of Saturn.

What if you could drive around Saturn's rings? Driving around the outside of Saturn's rings would take about a year.

Titan is the second-largest moon in the Solar System. It is larger than the planet Mercury.

Saturn and its rings would nearly fill the space between Earth and the Moon.

Jupiter, Uranus, and Neptune also have rings, but they are not very obvious. They have fewer rings and they are darker.

A spacecraft picture of part of Saturn's rings.

URANUS

Uranus is the seventh planet from the Sun. It is smaller than Saturn but still huge. It is a giant planet with a super thick atmosphere. You could not stand on Uranus.

Uranus's thick atmosphere is mostly made of hydrogen and helium, like Jupiter and Saturn. But it has more methane gas as well. The methane makes Uranus look blue.

Deep under the atmosphere, Uranus has hot fluids made of water, methane, and ammonia. When Uranus was forming, these fluids were ices. Because of this, Uranus is called an ice giant. Neptune is, too.

A spacecraft picture of Uranus.

Uranus has been visited by only one spacecraft. Voyager 2 flew by it in 1986.

Though much smaller than Jupiter and Saturn, Uranus is still much larger than Earth—63 Earths could fit inside it.

It takes Uranus 84 Earth years to go around the Sun. That means each season lasts 21 years. As it orbits the Sun, Uranus spins on its side. It is the only planet that spins on its side.

Uranus has a set of 13 mostly dark rings. It has at least 27 small moons.

Uranus is barely visible with just your eyes if you are somewhere very, very dark.

It would take 47.64 billion rolls of toilet paper (standard size, 2 ply) to reach Uranus from Earth.

The man who discovered Uranus wanted to name it after King George the Third. Can you imagine a planet named George? Uranus is named after the Greek god of the sky.

A 2006 Hubble Space Telescope picture of Uranus.

NEPTUNE

Neptune is the eighth planet from the Sun. It is just a little smaller than Uranus. Neptune is similar to Uranus in many ways. They are both known as ice giants. It also has a very, very thick atmosphere. You could not stand on Neptune.

Like the other giant planets, Neptune's thick atmosphere is mostly made of hydrogen and helium. But, like Uranus, it has methane gas as well. The methane makes Neptune look blue. Deep under the atmosphere, Neptune has hot fluids made of water, methane, and ammonia.

A 1989 spacecraft picture of Neptune including the Great Dark Spot, a storm that has disappeared since this picture was taken.

Like Uranus, Neptune has been visited by only one spacecraft. Voyager 2 flew by it in 1989.

Neptune takes almost 165 Earth years to go around the Sun.

Neptune has at least 13 moons. Most are small, except Triton. Triton is a world covered in strange ices.

You cannot see Neptune with just your eyes. It is too far away. You can see Neptune with a telescope.

Neptune is 30 times farther from the Sun than Earth. That causes Neptune and its moons to be very cold.

Neptune has the fastest winds of any planet in the Solar System. Scientists have measured wind speeds as high as 2,200 kilometers per hour (1,400 miles per hour).

Neptune is very far away. If the Sun were the size of a soccer ball, then Neptune would be about 7 football fields away.

DWARF PLANETS

Dwarf planets are like small planets. They orbit the Sun. They do not orbit something else like a moon does. They are big enough that gravity makes them round.

How is a dwarf planet different from an actual planet? A planet is much bigger than anything near it in its orbit. Dwarf planets have other stuff of similar size near their orbits.

Ceres is the largest asteroid. It is also a dwarf planet. It is the only asteroid big enough to be round like a ball, so it is the only asteroid called a dwarf planet. It is not called a planet because there are lots of other asteroids in similar orbits.

The other dwarf planets are out beyond Neptune. They are named Pluto, Eris, Makemake, and Haumea. They are all big enough to be round and are not moons. Pluto used to be called a planet, but then other objects

A spacecraft picture of the dwarf planet Pluto.

The first spacecraft to explore Ceres and Pluto reached them in 2015. Dawn went into orbit around Ceres. New Horizons flew by Pluto and its moons.

A spacecraft picture of the dwarf planet Ceres.

similar in size to Pluto were discovered. Eris is about the same size as Pluto. Instead of calling Eris and Makemake and Haumea planets, astronomers made a new name for these objects and called them dwarf planets.

Astronomers have discovered several objects that are probably dwarf planets. But they are so far away we can't tell if they are big enough to be round, so they can't yet be called dwarf planets.

Some of the dwarf planets have moons. Pluto has a big moon called Charon that is half as big as Pluto. Pluto also has at least four small moons.

Ceres was discovered in 1801. It was called a planet for a long time after its discovery until it was called an asteroid. All the other dwarf planets are bigger than Ceres.

Pluto takes 248 Earth years to go around the Sun. That is called one Pluto year. What was happening one Pluto year ago? The American Colonies belonged to Great Britain and the United States didn't exist.

Seven moons in the Solar System are bigger than Pluto.

WHO WAS THE PLANETARY SCIENTIST'S FAVORITE DISNEY CHARACTER?
PLUTO. 🙂

STUFF BEYOND NEPTUNE

There are lots of objects that are beyond distant Neptune but still orbit the Sun. These include the dwarf planets Pluto, Eris, Makemake, Haumea, and their moons. But that region also probably includes more than a trillion (1,000,000,000,000) objects. They are very hard to discover because they are so far away. We have discovered more than 2,000 so far.

Objects beyond Neptune are very cold. They get very little sunlight. They are often covered in different kinds of ice. One kind of icy object is called a comet. We'll talk about comets in the next chapter.

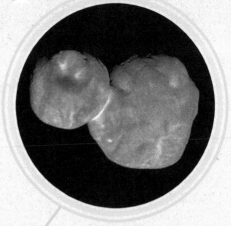

A spacecraft picture of the icy Trans-Neptunian Object 2014 MU69.

New Horizons is the only spacecraft to have visited objects beyond Neptune. It visited Pluto in 2015. In 2019, it flew by a city-sized icy body currently known as 2014 MU69. That was the most distant object visited by a spacecraft.

BIG WORD ALERT

Objects that spend most of their orbit beyond Neptune are called Trans-Neptunian Objects, or TNOs.

The object Sedna is so far from the Sun that it takes more than 11,000 years to go around the Sun.

CHAPTER 4

COMETS, METEORS, AND OTHER SKY SHOWS

Everything in space is moving. Sometimes when space stuff moves we can see cool shows in the sky. In this chapter we'll talk about what these things—including planets, comets, eclipses, *meteors*, and more—are. We'll also learn what they look like in the sky. Check out the Resources section at the back of the book (page 106) to find out when and where to look for these objects.

COMETS

Comets can be super cool looking. They are like big dirty snowballs, or snowy dirtballs. They can be the size of a city. Most spend most of their time beyond the orbit of Neptune. Out there, they are too small and far away for us to see.

Sometimes comets come in closer to the Sun and Earth. The Sun warms them. As they heat up, their ices turn into gases (just like dry ice does on Earth). As this happens, dirt and gas come off the comet. The large area of gas and dust is called the coma. The coma gets spread out by effects of the Sun into very, very long tails. Often comets have two tails. The dust tail looks white. The gas tail can be colorful.

A picture of a comet, including its tail.

BIG WORD ALERT

When a solid turns into a gas without becoming a liquid, it is called sublimation.

Unlike my dogs' tails, comet tails aren't always behind them. The Sun is causing the tails, so they always point away from the Sun. The tails are behind the comet when it is heading toward the Sun. But they are in front of the comet when it is heading away from the Sun.

Halley's comet is a famous comet. It comes near the Sun and Earth about every 76 years. Humans recorded seeing it more than 1,000 years ago. It will next be near Earth in 2061. How old will you be then?

Some comets take less than 100 years to go around the Sun. Some comets can take millions of years to go around the Sun.

Cartoons and movies show comets moving fast through the night sky. They do not do this. They move slowly through the sky from one night to the next.

A spacecraft picture of the rubber-duck shaped Comet 67P. Notice some hazy dust and gas coming off the comet.

METEORS, METEORITES, AND METEOR SHOWERS

Did you know that dust and small space rocks are hitting Earth's atmosphere all the time? That stuff can make neat streaks of light in the night sky. We call those meteors. Some people call them shooting stars or falling stars. They have nothing to do with stars.

Space dust hits Earth's atmosphere going really fast. The dust is going so fast that it burns up. The dust or rocks

A picture of meteors (the streaks of light) taken in an hour-long camera exposure during a meteor shower.

Every day about 100 tons of space dust and rocks enter Earth's atmosphere. That is more than the weight of 50 cars.

Meteor showers are named after the constellation where the meteors appear to begin. So Geminid meteors all seem to come from a point in the constellation Gemini.

glow as they burn up. That is what we see as a meteor.

A *meteorite* is a space rock that makes it to the ground without completely burning up. You can see meteorites in museums.

There are a few meteors every night. But some nights there are more meteors. We call this a meteor shower. You are more likely to see meteors during meteor showers. Meteor showers occur when Earth passes through the stuff left by a comet. The next year, Earth will pass through the stuff again on about the same date.

All you need to watch meteors are your eyes, no clouds, and patience. Go out, get comfortable, and stare at the sky. See the Resources section (page 106) to find out where you can learn about all the meteor showers of the year.

Two of the best meteor showers of the year are the Perseids, which peak around August 12 or 13, and the Geminids, which peak near December 13 or 14. During those showers, you can see 60 to 100 meteors per hour from a dark site. From a bright city or when there is a Full Moon, you will see fewer.

A meteor captured in a long camera exposure. The brightest star in the picture happens to be the North Star.

74

LUNAR ECLIPSES

You cast a shadow when you are in the light. Earth has a shadow. The Moon goes around Earth. Sometimes the Moon enters Earth's shadow. When it does, it is called a *lunar eclipse*. This happens when Earth is between the Moon and the Sun. During a lunar eclipse, the Moon gets dark.

A lunar eclipse will be visible from where you live only every few years on average. When one does occur, everyone on the night side of Earth can see it.

A picture of a total lunar eclipse. Notice the red color.

Unlike *solar eclipses* (page 76), which are dangerous to look at with your eyes, lunar eclipses are perfectly safe. See the Resources section (page 106) to learn how to find out when the next lunar eclipse will occur in your area.

Lunar eclipses often make the Moon look red. Red light can pass through the edges of Earth's atmosphere. Other colors don't make it all the way through the atmosphere. This is the same reason sunsets appear red.

Lunar eclipses usually last one to three hours. You can watch Earth's shadow gradually darken the Moon.

When the Moon enters only a part of Earth's shadow, it is called a *partial lunar eclipse*. When the Moon enters all of Earth's shadow, it is called a *total lunar eclipse*.

EARTH'S SHADOW

A picture of a partial lunar eclipse. The edge of Earth's shadow is round because Earth is round.

ORBIT OF THE MOON

SUN

ORBIT OF EARTH

COMETS, METEORS, AND OTHER SKY SHOWS

[Diagram Not to Scale]

SOLAR ECLIPSES

A solar eclipse occurs when the Moon's shadow crosses Earth. This happens when the Moon is between Earth and the Sun. That is the opposite of a lunar eclipse.

In a *total solar eclipse*, the Sun is completely blocked by the Moon. It gets almost as dark as night during a total solar eclipse. In a *partial solar eclipse*, the Sun is only partially blocked. Total *lunar* eclipses are visible from a very large area, but total *solar* eclipses are only visible from a small area. Unless you are very lucky, you will have to travel to see a total solar eclipse.

A picture of a total solar eclipse. The Moon is in front of the Sun.

A picture of a partial solar eclipse taken using a special safety filter. The Moon is partly in front of the Sun.

There is a funny word that means three objects lined up in space, like during a lunar or solar eclipse. That word is *syzygy*.

WARNING: DO NOT LOOK DIRECTLY AT THE SUN, EVEN DURING THE PARTIAL PHASES OF A SOLAR ECLIPSE. THE SUN WILL STILL BE SO BRIGHT IT CAN CAUSE BLINDNESS.

You can find out where to get information on when and where future eclipses will occur in the Resources section (page 106).

ORBIT OF THE MOON

ORBIT OF EARTH

MOON'S SHADOW

During a total solar eclipse, the Moon totally blocks the Sun for only about one to six minutes, depending on the eclipse.

SUN

[Diagram Not to Scale]

COMETS, METEORS, AND OTHER SKY SHOWS

MOON PHASES

When you look at the Moon, sometimes you see a whole circle. Sometimes you see only part of a circle. Why? Just like Earth, half of the Moon is always lit up by the Sun. That is daytime on the Moon. Just as on Earth, the part of the Moon in daytime changes. As the Moon orbits Earth, we see different amounts of the lit, or daytime, part of the Moon. We call these Moon phases or lunar phases.

Full Moon is when we see all the lit, or day, side of the Moon. See the picture on this page to learn the names of some of the other phases.

Lunar phases as seen from Earth (outer ring) and from above the North Pole (inner ring).

From one Full Moon to the next takes a little more than 29 days.

The Moon travels around Earth in about a month. The words *month* and *Moon* come from the same word origins.

At Full Moon, the Moon rises around the time of sunset.

AURORA

Auroras are beautiful light shows in the night sky. They are sometimes called the northern lights or southern lights.

Auroras occur only sometimes. They are hard to predict. Auroras can look like moving sheets or stripes of green, red, or other colored light.

Auroras occur near Earth's polar regions. In North America, auroras can usually be seen only in Alaska and Canada. Sometimes auroras are visible in other northern states. Auroras have also been seen on many other planets.

Auroras are caused by the Sun. Sometimes the Sun kind of burps. This is called a solar storm. The solar storm travels out from the Sun and hits Earth's atmosphere near the poles. That makes the pretty lights.

A picture of an aurora in Alaska.

The northern lights occur in the northern hemisphere. They are also called aurora borealis.

The color of an aurora is related to the gas the solar storm hits in Earth's atmosphere. For example, green auroras are produced by oxygen.

The southern lights occur in the southern hemisphere. They are called aurora australis.

COMETS, METEORS, AND OTHER SKY SHOWS

PLANETS

Several planets are bright and easy to see in the night sky. Mercury, Venus, Mars, Jupiter, and Saturn look like bright stars. They are so far away they look like dots, just like the stars. When we look at a planet, we are seeing sunlight bouncing off the planet.

Planets move a little from one night to the next. Sometimes they cannot be seen at all. To find them, you need to learn where and when to look. The Resources section of this book (page 106) lists materials that tell you where to find the planets.

A picture of the Moon with Jupiter (bright dot on the left) and Mars (faint dot between Jupiter and the Moon).

The word *planet* comes from the Greek word meaning "wanderer." That is because the planets appear to wander through the sky.

Venus is the brightest planet in the sky. Jupiter is the next brightest. They are both brighter than the brightest star in the night sky (Sirius).

The distance between Earth and Mars changes a lot. That is because of their orbits. When Mars is close to Earth, it is brighter than the brightest star. When Mars is far from Earth, it looks like a normal star.

ROCKETS, SATELLITES, ASTRONAUTS, AND MORE!

Sending objects into space is exciting! But it is also hard. We use rockets to get off Earth and into space. We use robotic spacecraft for exploration around Earth and far out in space. People travel to space, and a few are living in space right now. Here, we'll learn about all that, and about your future in space.

ROCKETS

We use rockets to get spacecraft to space. There are many types of rockets. But the rockets that go to space are usually long, thin, and round.

Rocket engines burn fuel and spit hot gases out the back of the rocket. That pushes the rocket forward. It's like when you blow up a balloon and let go without tying it. The air rushes out the back and pushes the balloon forward.

Why can't we use airplanes to go to space? Because there is no air in space. Airplane wings use air to hold the airplane up. Airplane engines use air, too.

Apollo 11 Saturn V rocket launch.

Some rockets burn liquid fuel. Some rockets burn solid fuel. Solid fuel is also used in model rockets.

A rocket takes about eight minutes to get a spacecraft into orbit in space.

The space shuttle was a rocket that carried what looked like a plane on it. When the shuttle came back from space, it flew using wings when it got close enough to the ground. But it still had to use rockets to get to space.

Smaller rocket engines are often used on spacecraft to help get them to their destination.

The Saturn V rocket that launched Apollo 12 was struck twice by lightning in the first minute after liftoff, but that did not keep the mission from being successful.

The Saturn V (pronounced "Saturn five") was the world's most powerful rocket. It was used to send astronauts to the Moon.

WHAT IS AN ASTRONAUT'S FAVORITE MEAL?

LAUNCH. 🙂

The last space shuttle launch (2011).

SATELLITES

A satellite is a spacecraft that is orbiting Earth. There are thousands of satellites traveling around Earth right now.

You may have used one of those satellites without even knowing it. Some of them are used to send TV signals. Look at the roofs of houses near you. You will probably see some have satellite dishes. Those dishes receive signals from TV satellites thousands of kilometers (or miles) above Earth.

A satellite just after release from the space shuttle. The space shuttle's tail can be seen behind the satellite.

The first satellite was Sputnik, launched in 1957.

Satellites like the International Space Station orbit Earth in just an hour and a half (90 minutes), or about the time it takes you to watch a movie.

WHAT DID THE ASTRONAUT USE TO SEE IN THE DARK?

A SATEL-LITE. 🙂

Weather satellites take pictures of Earth from space. Scientists use the pictures to help predict the weather.

Other satellites are used to help people know where they are on the surface of Earth. Electronic maps in cars and phones use the positions of GPS satellites to figure out where you are.

To stay in space, a satellite has to move fast to not fall back to Earth. To orbit Earth, a satellite has to travel about 28,000 kilometers (17,500 miles) per hour. That is why we have to use rockets to make them go fast.

The oldest satellite still in space is the US Vanguard 1. It was launched in 1958.

BIG WORD ALERT

Technically, a satellite is anything that orbits, or goes around, another body. Moons are called *natural satellites*. Spacecraft that go around another body are called *artificial satellites*.

ROCKETS, SATELLITES, ASTRONAUTS, AND MORE!

FLYBY SPACECRAFT

The easiest way to visit another planet with a spacecraft is to fly by and just keep going. That is how we first explored most places we visited in the Solar System. In fact, every planet in our Solar System was first explored by a spacecraft that flew by and kept going. (We use the term *flyby*, but remember, no actual flying happens. No air is involved.)

Why fly by? Because it is much easier than slowing the spacecraft down to go into orbit.

After exploring with *flyby spacecraft*, sometimes we go back with orbiters. Then we sometimes use atmospheric probes, landers, or even rovers that drive around on the surface.

The New Horizons spacecraft that flew by Pluto. The radio dish is used to communicate with Earth.

The Voyager 2 spacecraft flew by all four giant planets. It is still the only spacecraft to have visited Uranus or Neptune.

The first planet besides Earth visited by a spacecraft was Venus. The US Mariner 2 spacecraft flew by Venus in 1962.

Voyager 1 is the farthest spacecraft from Earth.

ORBITERS

An orbiter is a spacecraft that orbits an object. In other words, it keeps going around that object. To put a spacecraft in orbit around another planet or moon, the spacecraft has to be slowed down. Otherwise it will fly by.

Flyby spacecraft get only one brief look at a planet or moon. Orbiters can keep looking and studying the planet, moon, or other object they are orbiting. That allows us to learn a lot more. Orbiters can also take lots of pictures and create maps.

An artist's painting of the Mars Odyssey spacecraft orbiting over Mars's icy South Pole.

WHAT WAS THE ORBITER'S FAVORITE PLAYGROUND EQUIPMENT?

THE MERRY-GO-ROUND. ☺

Some spacecraft can keep working a long time. The Mars Odyssey orbiter has been working at Mars since 2001.

The Vanguard 1 satellite launched in 1958. It is an Earth orbiter. It stopped working in 1964 but has stayed in orbit. It has gone around Earth more than 200,000 times.

Spacecraft have orbited all the planets except Uranus and Neptune.

ROCKETS, SATELLITES, ASTRONAUTS, AND MORE!

ATMOSPHERIC PROBES

The next step after sending an orbiter to a planet or other object is to send a probe to enter its atmosphere or land on the object.

The friction from hitting the gases in the atmosphere at high speed heats up the probe. (Friction is what warms your hands up when you rub them together fast.) Probes use a heat shield to protect them from the heat.

After it slows down, the probe drops its heat shield to expose scientific instruments. Then it floats down on a parachute. We use probes to learn more about atmospheres.

As the Jupiter Galileo probe got deeper in Jupiter's atmosphere, it eventually melted and got crushed by the high temperatures and pressures.

An artist's painting of the Galileo probe in Jupiter's atmosphere just after dropping the heat shield.

Venus atmospheric probes named Vega 1 and Vega 2 went one step further. These probes used balloons that floated in the Venus atmosphere.

The Huygens probe was designed as an atmospheric probe to study the thick atmosphere of Saturn's moon Titan. Huygens took more than two hours to float down on a parachute through the Titan atmosphere.

LANDERS

We have landed spacecraft on planets and moons that have a solid surface. These spacecraft are often called landers because they land.

Robotic landers have landed successfully on the Moon, Venus, Mars, and some other smaller objects. Humans have only landed on the Moon.

The first successful soft landing on an object in space was the Soviet Luna 9 spacecraft on the Moon in 1966. Others before that crashed either accidentally or on purpose.

The Phoenix Mars lander used a robotic arm to dig into the surface, where it found water ice under the dirt.

On Mars, some landers have landed using airbags kind of like those used in cars.

The Surveyor 3 robotic lander being visited by Apollo 12 astronaut Pete Conrad.

Continued >

ROCKETS, SATELLITES, ASTRONAUTS, AND MORE!

LANDERS

Spacecraft are going very fast, and they need to slow down to land. On places like the Moon that have no atmosphere, the spacecraft uses rockets to slow down. On places with atmospheres, the spacecraft has to start the landing like an atmospheric probe. It uses a heat shield, then parachutes. Then it often uses rockets right before landing.

The view of the Mars surface from the Viking 2 lander.

Landers that have gone to Venus lasted only an hour or two on the surface. That is because of the high temperatures and pressures that quickly destroy parts of the lander.

We've also landed successfully on a comet, an asteroid, and Saturn's moon Titan.

ROVERS

Landers that have wheels and are designed to drive are called rovers. Rovers have been successfully used on the Moon and Mars.

The Soviet Union and China have used robotic rovers on the Moon. Americans brought rovers with them on the last three Apollo missions to the Moon. Astronauts drove those rovers.

NASA has had successful robotic rovers on Mars. These rovers, like other spacecraft, are designed to last for years. No one is there to fix a rover or change a tire on Mars.

The lunar rover of Apollo 17 on the Moon. Astronaut Gene Cernan is driving.

The Opportunity rover has driven the farthest on another object. In its 14 years working on Mars, it drove about 45 kilometers (28 miles). That is a little longer than a marathon race.

Apollo 17 astronauts fixed a broken fender on their rover using maps and duct tape.

The two Mars exploration rovers each have LEGO bricks made of aluminum on their landers.

WHAT DID THE ASTRONAUT NAME HIS DOG?

ROVER. 🙂

GETTING PEOPLE TO SPACE

A few people are living in space right now. But getting people to space and having them live there is hard. Unlike robotic spacecraft, people require food, air, and other things. And the rockets have to be gentle on the human body.

But people can and do live in space. We call those people astronauts. Russian astronauts are called cosmonauts.

To provide what astronauts need to live in space also requires more room. Their spacecraft have to be bigger than robotic spacecraft. That way they can have air and room to move, and they have room to store their food and equipment.

When astronauts leave an air-filled spacecraft, they have to wear special clothing. This clothing is called a spacesuit. It is filled with air. Astronauts

Astronaut John Glenn with the Mercury spacecraft he used to become the first American to orbit Earth in 1962.

NASA's early one-person flights were on what were called Mercury spacecraft. Then they had two-person spacecraft called Gemini. Then they moved to three-person spacecraft called Apollo.

can breathe inside a spacesuit for a few hours. When astronauts go outside a spacecraft, it is sometimes called a spacewalk, even though there is nothing to walk on.

People first went to space in the early 1960s. In the beginning, the smallest possible spacecraft were used to send one person to space. Later, anywhere from three to seven people could be sent at one time. And larger spacecraft were built for people to live in for long periods of time. Those are called space stations.

The original spacecraft did not have bathrooms. Just bags.

The first person to go into space was Soviet cosmonaut Yuri Gagarin in 1961.

The space shuttle could carry as many as seven astronauts to space.

The first American spacewalk by Ed White in 1965.

WHAT WAS THE SPACE TRAVELER'S FAVORITE FOOD?

ASTRONUTS. 🙂

93

HUMANS TO THE MOON

Almost all the people who have gone to space have stayed near Earth. Most have been in Earth orbit. But a few NASA missions in the 1960s and '70s sent humans to the Moon. They were part of the Apollo program. The Moon is the only place besides Earth that people have visited.

Getting humans safely to the Moon is much harder than putting them in Earth orbit. The Moon is much farther away. It took the astronauts about three days to get to the Moon. Special spacecraft had to be designed to get to the Moon and to land on it.

Three astronauts went on each mission to the Moon. One would stay in orbit around the Moon. The other two would land on the surface. They collected rocks and dirt to bring back to Earth for scientific study. They also put scientific instruments on the Moon.

The Apollo 11 Lunar Module on the Moon. Astronaut Buzz Aldrin can be seen in front of the lander.

The Apollo missions brought back about 382 kilograms (842 pounds) of moon rocks and dirt to Earth. That is about the mass (or weight) of five adults.

We learned a lot from the rocks they brought back. One of the biggest discoveries was that the Moon formed when a planet-sized body crashed into Earth 4.5 billion years ago. The Moon formed from material that got thrown out from Earth in the impact.

Only 12 people have walked on the Moon. They did it two at a time.

The Apollo 10 spacecraft were named after *Peanuts* cartoon characters. The command service module that carried astronauts to and from the Moon was Charlie Brown. The lunar module that went down near the Moon's surface was Snoopy.

Apollo 17 astronaut Harrison Schmitt on the Moon. Because there is no atmosphere, the sky is black during the day.

WHY DIDN'T THE ASTRONAUT LIKE THE RESTAURANT ON THE MOON?

THERE WAS NO ATMOSPHERE. ☺

SPACE STATIONS

A space station is like a home in space for astronauts. Astronauts travel to space in another spacecraft and then connect to the space station. They live on board the space station, then eventually come home using another spacecraft.

The Russian Mir space station as seen from a US space shuttle in 1997.

Most space stations have more than one docking port. That way, more than one spacecraft can be attached to the space station at the same time.

Except for the International Space Station and Tiangong-2, all other space stations have reentered Earth's atmosphere and burned up.

The first space stations were launched into Earth orbit in the 1970s. They were small. These included several Soviet Salyut space stations and the American Skylab. Later, the Soviets built the Mir space station, which was bigger. It was made of several separate pieces. More recently, several countries worked together to create the very large International Space Station. The Chinese also have launched their smaller Tiangong space stations.

Space stations allow astronauts to stay in space longer. They can carry out more scientific experiments with the extra time.

The Skylab space station was damaged during launch. But the first astronauts to visit Skylab were able to go outside the space station and make repairs.

The four longest stays in a space station were all a year or more on board the Mir space station.

WHERE DID THE ASTRONAUT PARK HIS SPACE STATION?

IN A PARKING SPACE. 🙂

INTERNATIONAL SPACE STATION

The International Space Station (ISS) is the largest space station that has ever been built. The first piece was launched in 1998. Many more pieces were added over many years. It is still in orbit.

Astronauts usually spend a few months on board the ISS. They come from many different countries. Three to six astronauts usually live on board. There has been someone living there constantly since the year 2000. Astronauts do a lot of science experiments on the ISS.

A 2010 picture of the International Space Station taken from the Space Shuttle.

The ISS has two bathrooms on board.

The ISS is about the size of a football field.

WHAT WAS THE ASTRONAUT'S FAVORITE KEY ON THE KEYBOARD?

THE SPACE BAR. ☺

Some spacecraft carry astronauts to the ISS. Some carry cargo like food and new science experiments.

The ISS has very large solar panels that provide electricity using the light from the Sun. They are similar to solar panels used on Earth to produce electricity.

You can see the ISS at night. It looks like a very bright star that is moving across the sky. We can see it because of the sunlight reflecting off it. See the Resources section (page 106) to find out where and when to look for it.

The ISS orbits only about 400 kilometers (about 250 miles) up. If you could drive a car straight up, it would take you about four hours to reach the ISS. Of course, you would have trouble trying to drive a car straight up.

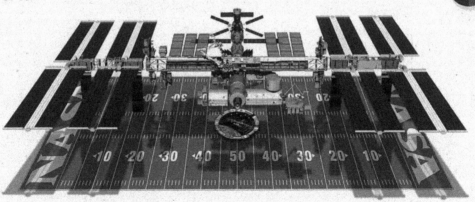

The size of the International Space Station compared to a football field.

ROCKETS, SATELLITES, ASTRONAUTS, AND MORE!

ASTRONAUTS

Astronauts are people who go to space. Their job is hard and dangerous. But it is also fun and amazing. They get to experience an incredible view of Earth from space. They also get to float in space.

Astronauts have to study and train a lot before they go on a space mission.

When a rocket takes astronauts into space, the rocket shakes and the astronauts get pushed back in their seats. When they are in space, it is like there is no gravity. They float. Their tools float. Their food floats. Sometimes they demonstrate what happens when things float in space by playing with toys in space.

Astronauts also get shaken around when their spacecraft returns to Earth. The spacecraft has to slow down a lot before they can land.

100

The 13 people who were on the ISS when a space shuttle visited in 2010. Astronauts can be "upside down" because everyone is floating.

Fewer than 600 people have flown in space.

Over five space missions, Gennady Padalka spent a total of almost 2.5 years in space.

A lot of different people become astronauts. For example, some are pilots, some are scientists or engineers, and some are medical doctors.

Astronauts have to be brave. Going to space is dangerous. Some astronauts have died. But what astronauts do is important, and they inspire the rest of us to do great things.

The longest time spent in space on one spaceflight by a human is almost 438 days. That is more than one year and two months. The record was set by Valeri Polyakov on the Mir space station.

Astronauts sometimes train underwater to get used to the feeling of floating in space.

Astronaut Marsha Ivins in space. An astronaut's hair can float in space.

ROCKETS, SATELLITES, ASTRONAUTS, AND MORE!

YOUR SPACE FUTURE

Do you think space is interesting? There are lots of ways for you to enjoy space more. You can read about it in books and on the Internet, or watch videos about it. The Resources section (page 106) lists ways you can learn more about space.

You can also teach people about space. You can share super cool space facts from this book with your friends and family. You can talk with other people about space topics that you enjoy.

You can learn about the night sky. You can learn how to find cool things in the night sky and show your friends. I have another book that will teach you how to do that. It is called *Astronomy for Kids: How to Explore Outer Space with Binoculars, a Telescope, or Just Your Eyes!*

When you get older, you can get a space job if you want. There are lots of jobs that have to do with space. There are astronauts and scientists and engineers. But there are also writers and artists and managers and many more jobs.

Or you can just do space things as a hobby. Whatever you choose to do, I hope you will continue to enjoy learning about the amazing wonders of space. Space facts equal space fun!

GLOSSARY

ASTERISM: a pattern of stars in the sky forming a recognizable shape.

ASTEROID: a small rocky or metallic object in space.

ATMOSPHERE: the gases (for example, oxygen and nitrogen) held by gravity to a planet, moon, or other body.

COMET: icy dirtballs (or dirty iceballs) in space that can be the size of a city; when they come near the Sun, dust and gas come off the surface and form fuzzy-looking objects with tails.

CONSTELLATION: one of 88 internationally agreed-upon patterns of stars; each constellation is given boundaries, dividing the total sky into 88 areas.

DWARF PLANET: a round body that orbits the Sun but not another body. Unlike a planet, it has objects of similar size in its orbit. It is typically much smaller than a planet.

FLYBY SPACECRAFT: a spacecraft that flies by an object (like a planet) without going into orbit or trying to land.

GALAXY: a collection of millions, billions, or even trillions of stars, as well as dust and gas, all held together in one group by gravity. We live in the Milky Way Galaxy.

LANDER: a spacecraft designed to land on another body like a planet or moon.

LUNAR ECLIPSE: an event when the Moon is exactly on the opposite side of Earth from the Sun, causing the Moon to enter Earth's shadow.

MASS: a property of a physical body often measured in grams or kilograms. Mass is like weight but is independent of gravity. So you would weigh less on the Moon because it has lower gravity than Earth. But your mass would stay the same.

METEOR: a streak of light in the sky caused by space stuff burning up as it hits the upper atmosphere of Earth at very high speeds.

METEORITE: a rock that makes it through the atmosphere to the surface of Earth.

MOON: an object that orbits a planetary body; moons are often called natural satellites. Earth's moon is known as the Moon.

NEBULA: a giant collection of dust and gas in space; some nebulae are left over from material blown outward during the end stages of stars, while others are star-forming regions.

NEUTRON STAR: what is left behind after really big stars explode. Neutron stars are very small. They are about the size of a city. But they still have the mass of a star.

NUCLEAR FUSION: when two nuclei get squished together to form a heavier nucleus. For example, stars combine hydrogen into helium through nuclear fusion. This releases a huge amount of energy.

ORBIT: to go around another body, like Earth orbits the Sun. Also, the path a planet or moon or other object follows as it goes around another object; for example, Earth's orbit around the Sun is approximately a circle.

ORBITER: a spacecraft that orbits a body like a planet or moon.

PLANET: originally, one of the five "wandering" star-like objects in the night sky (Mercury, Venus, Mars, Jupiter, and Saturn) that move relative to the stars. Since the telescope was invented, the definition has varied. Now it is an object that orbits the Sun, that is rounded by gravity, and that does not have any objects of similar size near its orbit.

PLASMA: a state of matter, kind of like an electrical charge–filled gas.

RED GIANT: one of the end phases of the life of a middle- to low-mass star in which the star expands (becomes giant) and as a result, its surface cools, causing the red color.

ROVER: a lander that has wheels and can drive on an object in space.

SOLAR ECLIPSE: an event that occurs when the Moon is exactly between the Sun and Earth, causing the Moon's shadow to cross part of Earth.

SOLAR SYSTEM, THE: the Sun and all the planets, moons, and other stuff that orbit the Sun.

SPACE: sometimes called outer space, space is all the stuff beyond Earth and its atmosphere.

SPACECRAFT: a vehicle or machine designed to fly in space.

SPECTROSCOPY: using the detailed colors of light to learn what things are made of.

STAR: an enormous ball of hot glowing gas (technically plasma). The Sun is a star. All stars except the Sun are so far away that they appear as dots of light in the night sky.

STAR CLUSTER: a group of stars that not only appear near one another in the sky but are also actually near one another in space.

SUPERNOVA: the explosion of a huge star at the end of its normal life, causing brightening over weeks and months and years.

UNIVERSE: all of space and everything in it. That includes Earth and the farthest stars from Earth and everything in between.

WHITE DWARF: what is left at the end of the lives of most stars. It forms when the star runs out of fuel to burn. White dwarfs are much smaller than stars but have similar mass.

RESOURCES

GENERAL SPACE EXPLORATION INFORMATION AND FUN

Check out the author's other children's space books, including *Astronomy for Kids: How to Explore Outer Space with Binoculars, a Telescope, or Just Your Eyes!*

RandomSpaceFact.com: Bruce Betts's website offers information about the author and links to his other astronomy-related content, including videos, a radio show, classes, and social media accounts.

The Planetary Society's *Random Space Fact with Dr. Bruce Betts*: Planetary.org/rsf. This website offers fun, humorous short videos full of space facts. Great for kids!

The Planetary Society: Planetary.org. Visit this website to get space exploration updates and to join the world's largest space interest group.

NASA Space Place: SpacePlace.NASA.gov. This website, designed for kids, lets visitors learn about space and enjoy space-related activities.

NASA Kids Club: NASA.gov/kidsclub/index.html. Visit this page for games and information about NASA.

WHERE TO LOOK FOR PLANETS, SKY CHARTS, AND MORE

Sky and Telescope magazine's website: SkyAndTelescope.com. In particular, check out "This Week's Sky at a Glance," including where to look for planets. Look on the menu under Interactive Tools to find a sky chart creator, Jupiter moon position information, and other useful tools. You can also find telescope buying guide information. You can also subscribe to their magazine, which includes upcoming sky information and star charts.

Astronomy magazine's website: Astronomy.com. In particular, check out "Sky This Week," which includes where to look for planets. You can also subscribe to their magazine, which includes upcoming sky information and star charts.

Planetary Radio: Planetary.org/radio. Listen to the end of this weekly podcast/radio show for night sky info from the author of this book. He'll tell you where to look for planets, as well as give you space facts and a trivia contest.

Stellarium (free night sky software for PC, Mac, or Linux): Stellarium.org. This and similar software will show you what the night sky will look like on a given night. Make sure you set your city as your default location when you use it. Many other paid software programs also exist and can be purchased from Amazon or astronomy websites.

Night sky apps for iOS and Android can be found by searching the appropriate app store. An app based on Stellarium is available for a small purchase price. These apps will show you what the night sky will look like on a given night.

INFORMATION ABOUT THE NEXT ECLIPSES

NASA's eclipse webpages: Eclipse.GSFC.NASA.gov/eclipse.html

The Planetary Society: Your guide to future total solar eclipses. Here's an article about all total solar eclipses until 2030: Planetary.org/eclipse

METEOR SHOWER INFORMATION

American Meteor Society Meteor Shower Calendar: AMSMeteors.org /meteor-showers/meteor-shower-calendar

Astronomy Calendar of Events for the current year (includes all types of celestial events): SeaSky.org/astronomy/astronomy-calendar-current.html

INTERNATIONAL SPACE STATION FLYOVERS

NASA's Spot the Station (make sure you enter your city!): SpotTheStation. NASA.gov

Heavens Above (make sure you enter your city!): Heavens-Above.com. The site also has predictions for other satellites, sky charts, and other information.

FIND YOUR LOCAL ASTRONOMY CLUB OR ORGANIZATION

NASA/JPL Night Sky Network list of clubs and events: NightSky.JPL.NASA.gov /club-map.cfm

Sky and Telescope: SkyAndTelescope.com/astronomy-clubs-organizations

The Astronomical League: www.AstroLeague.org/astronomy-clubs-usa-state

INDEX

ACKNOWLEDGMENTS

Thanks to Jennifer Vaughn for her wise professional guidance and her magnificent love and support. Thanks to my sons, Daniel and Kevin Betts, for their support of projects like this as well as bringing happiness and fulfillment to my life. I am grateful to my parents, Bert A. and Barbara Lang Betts, for supporting my early interest in space and education. Thanks to Kathleen Reagan Betts for being such a great mom and for all the children's books she read to our sons. And thank you to Bill Nye and all the staff and supporters of The Planetary Society for their interest in and support of my broader science and education efforts.

Thanks to my editor, Susan Randol, for all her suggestions and comments that have made this a better book, and for being a pleasure to work with. My appreciation goes out to Merideth Harte, Vanessa Putt, Andrew Yackira, Sara Feinstein, and the rest of the Callisto Media team for their positive attitudes and professional efforts that enabled this book and made it better.

Finally, thanks to all the readers of this book for giving it a try and for enjoying the wonders of space. Enjoy!

ABOUT THE AUTHOR

DR. BRUCE BETTS is a planetary scientist and children's book author who loves teaching people about planets, space, and the night sky in fun and entertaining ways. He has lots of college degrees, lots of big dogs, and two sons.

Dr. Betts is the chief scientist and LightSail program manager for the world's largest space interest group, The Planetary Society. He has a BS in physics and math and an MS in applied physics with an emphasis in astronomy from Stanford University, as well as a PhD in planetary science with a minor in geology from Caltech. His research there and at the Planetary Science Institute focused on infrared studies of planetary surfaces. He managed planetary instrument development programs at NASA headquarters.

He is the author of *Astronomy for Kids: How to Explore Outer Space with Binoculars, a Telescope, or Just Your Eyes!* and *V.R. Space Explorers: Titan's Black Cat.*

At The Planetary Society, he heads the Science and Technology and the Education and Outreach programs. He has managed several flight hardware projects and has led additional science and outreach projects. He is the program manager for the Society's largest project: the LightSail solar sail spacecraft. He regularly writes for the member magazine *The Planetary Report* and his blog at Planetary.org. His popular Twitter feed (@RandomSpaceFact) and Facebook page (Facebook.com/DrBruceBetts) provide easy night sky astronomy and random space facts. His *Random Space Fact* video series (Planetary.org/rsf) provides space facts mixed with humor and graphics. He also co-hosts the "What's Up?" feature on the weekly *Planetary Radio* show (Planetary.org/radio, 100+ stations, XM/Sirius, podcast). He has been a guest expert on History Channel's *The Universe*, is a frequent contributor to *Professional Pilot* magazine, and has appeared frequently in TV, print, Web media, and public lectures. Dr. Betts is an adjunct professor with California State University, Dominguez Hills, and his Introduction to Planetary Science and Astronomy class, featuring lots of pretty space pictures, is available free online (Planetary.org/bettsclass). He is an alumnus senior scientist with the Planetary Science Institute. His website is RandomSpaceFact.com.

CPSIA information can be obtained
at www.ICGtesting.com
Printed in the USA
JSHW010855050820
7041JS00005B/4